PERENNIAL MATH

CONTEST PROBLEMS

2014-2017

Dr. Sylvia Dean has been a gifted specialist for 25 years in Huntsville, Alabama. Dr. Dean created Perennial Math in 2011 upon completion of her doctoral degree. She has hosted many robotics, technology, and mathematics competitions throughout the years and has tirelessly worked on providing educational opportunities to gifted students around the world. It has always been her ambition to leave students with the feeling that "math is cool."

The Rookie Level was added to Perennial Math in 2012. These problems are challenging and students should know up front that scoring a 3 out of 6 on any of these tests is extraordinary and above average. Great problem-solvers learn persistence and appreciation for challenge. Good luck as you encourage your students to take the quest to become great mathletes.

ROOKIE LEVEL

Grades 3-4

Name________________________________

PERENNIAL MATH

November 2014 Rookie Level

TEST # 1

1) John has 18 tools on his workbench. He has twice as many screwdrivers as hammers. How many hammers are on his workbench?

2) Andrew buys an eraser for 30 cents. He pays for it with 2 quarters. How many different ways can Andrew get change back?

3) Billie can put together a puzzle in 20 seconds. Charlotte takes 45 seconds to put together the same puzzle. After 3 minutes, how many more puzzles does Billie have than Charlotte?

4) A group of 48 teachers is taking 11 cars to a workshop. If each car has either 4 or 5 teachers, what is the number of cars carrying 5 teachers?

5) Use the balancing scales to determine how much the two turkeys weigh.

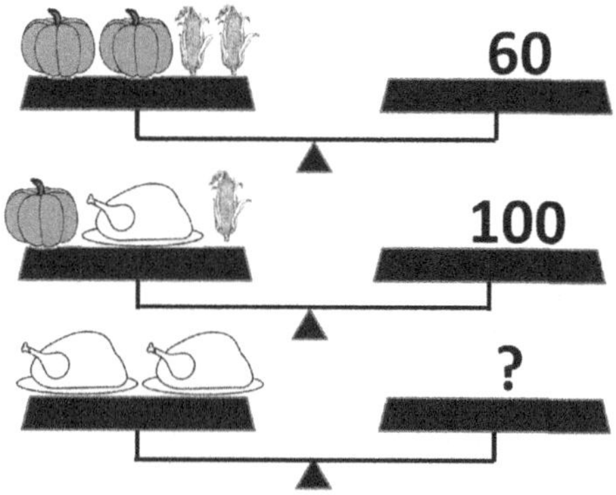

6) Maxine told Greg to pick a number, add 8 to it, double that sum, and then divide by 4. Greg said his answer was 10. What number did he start with?

PERENNIAL MATH

TEST #1

November 2014 Rookie Level Answers

1) John has 18 tools on his workbench. He has twice as many screwdrivers as hammers. How many hammers are on his workbench?	6
2) Andrew buys an eraser for 30 cents. He pays for it with 2 quarters. How many different ways can Andrew get change back?	9
3) Billie can put together a puzzle in 20 seconds. Charlotte takes 45 seconds to put together the same puzzle. After 3 minutes, how many more puzzles does Billie have than Charlotte?	5
4) A group of 48 teachers is taking 11 cars to a workshop. If each car has either 4 or 5 teachers, what is the number of cars carrying 5 teachers?	4
5) Use the balancing scales to determine how much the two turkeys weigh. 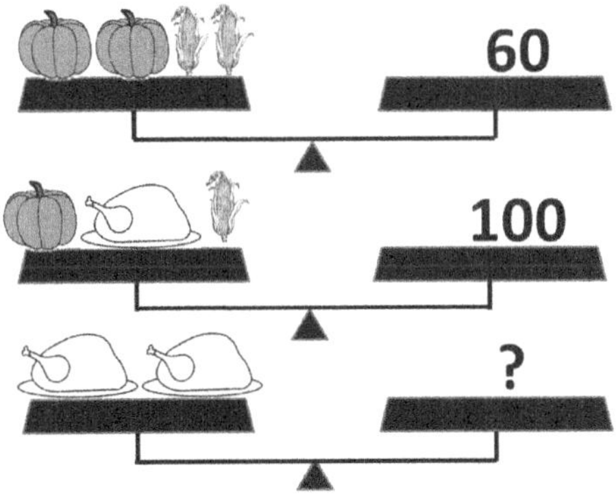	140
6) Maxine told Greg to pick a number, add 8 to it, double that sum, and then divide by 4. Greg said his answer was 10. What number did he start with?	12

PERENNIAL MATH

TEST #1

November 2014 Rookie Level Solutions

We are only providing some suggested solutions below. There are always multiple methods to solve problems.

1) Students can do guess and check or make a chart. There are 6 hammers on the workbench.

Hammers	Screwdrivers (2 x Hammers)	Total Tools (18)
1	2	3
2	4	6
3	6	9
4	8	12
5	10	15
6	12	18

6

2) Make an ordered list. Start with one type of coin, use the most of that coin until you have used it up and proceed to the next coins using the same process. There are nine ways to make change.

Dimes	Nickels	Pennies
2		
1	2	
1	1	5
1	0	10
0	4	0
0	3	5
0	2	10
0	1	15
0	0	20

9

3) Billie can put together a puzzle in 20 seconds. So convert this to how many she can do in 3 minutes. Three minutes equal 180 seconds, so she can do 180/20 = 9 puzzles. Charlotte takes 45 seconds to put together the same puzzle. 180/45 = 4 puzzles. Nine puzzles minus four puzzles equals five puzzles more.

5

PERENNIAL MATH TEST # 1

November 2014 Rookie Level Solutions

4) Make a chart to add up how many of each vehicle type and how many total teachers. There are four cars that have 5 teachers in them.

4 passenger cars	1	2	3	4	5	6	7
5 passenger cars	10	9	8	7	6	5	**4**
Total cars (11)	11	11	11	11	11	11	11
Total teachers (48)	54	53	52	51	50	49	48

4

5) Use logic. Two pumpkins plus two corn = 60. Therefore, half of this (1 pumpkin and 1 corn) = 30. The next equation has (1 pumpkin and 1 corn) plus one turkey = 100. Substitute 30 in for the (1 pumpkin and 1 corn) and the turkey = 70. Now you can solve for 2 turkeys. 2 x (70) = 140

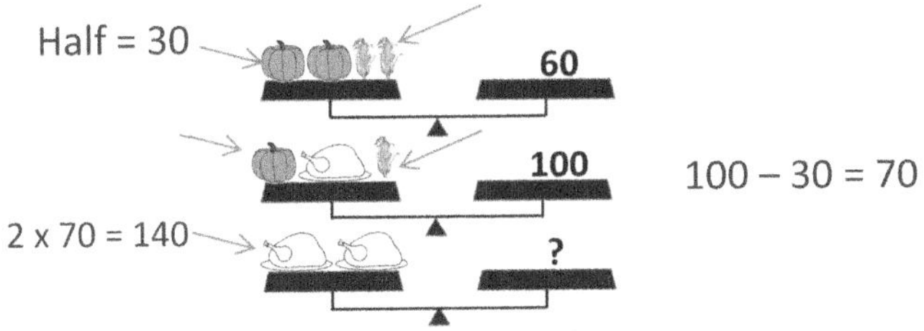

140

6) <u>Method 1</u>

"X" will represent the number Greg picked. We need to add 8 to it and then multiply it by 2 and then divide it by 4 and the answer is 10.

So, let's solve for X:

$$\frac{2\,(X + 8)}{4} = 10 \qquad \frac{2\,(12 + 8)}{4} = 10$$

solving the equation, X = 12

<u>Method 2</u>

Work backwards - Start with 10 and multiply by 4 which is 40. Divide this by 2 and you get 20. Subtract 8 and you will have 12. You can check your work now going forward.

12

Name________________________________

PERENNIAL MATH

TEST # 2

December 2014 Rookie Level

1) Complete the pattern. What is the sum of the three missing numbers? 2, 7, 5, 10, 8, 13, 11, ___ , ___ , ___

2) Jenny is ordering cans of waffle mix and they costs $5.19 per can or a 3-pack costs $13.57. She must also pay $8 for shipping. She has a $100 gift card. What is the most number of cans she can buy?

3) In 3 minutes Marvin can saw a log into two pieces. If he cuts at the same rate, how many minutes will it take him to saw the log into twelve pieces?

4) Larry the elf was counting toys in Santa's Shop. He saw that there were some tricycles and some bicycles. He counted 21 seats and 48 wheels. How many bicycles did he count?

5) If 1 square = 3 stars and 1 triangle = 7 stars, and 1 circle = 2 stars, how many stars would need to be added to the right side of the scale to balance the scale?

6) The Johnson family members each brought a $10 gift for everyone in the family. If there are 10 people in the family how many total dollars were spent on gifts?

PERENNIAL MATH

TEST #2

December 2014 Rookie Level Answers

1) Complete the pattern. What is the sum of the three missing numbers? 2, 7, 5, 10, 8, 13, 11, ___ , ___ , ___	49
2) Jenny is ordering cans of waffle mix and they costs $5.19 per can or a 3-pack costs $13.57. She must also pay $8 for shipping. She has a $100 gift card. What is the most number of cans she can buy?	20
3) In 3 minutes Marvin can saw a log into two pieces. If he cuts at the same rate, how many minutes will it take him to saw the log into twelve pieces?	33
4) Larry the elf was counting toys in Santa's Shop. He saw that there were some tricycles and some bicycles. He counted 21 seats and 48 wheels. How many bicycles did he count?	15
5) If 1 square = 3 stars and 1 triangle = 7 stars, and 1 circle = 2 stars, how many stars would need to be added to the right side of the scale to balance the scale? 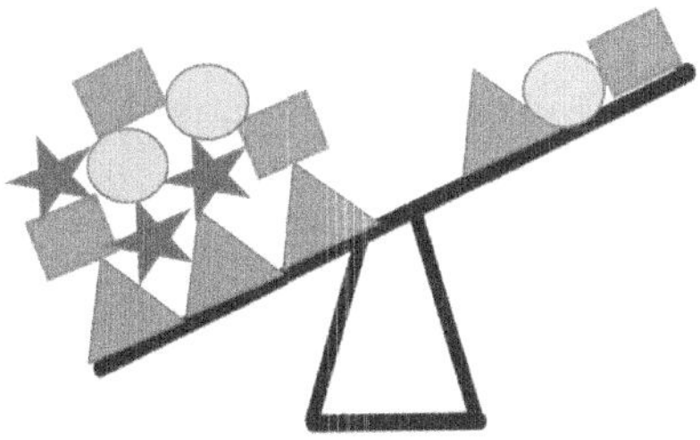	25
6) The Johnson family members each brought a $10 gift for everyone in the family. If there are 10 people in the family how many total dollars were spent on gifts?	900

PERENNIAL MATH

TEST # 2

December 2014 Rookie Level Solutions

We are only providing some suggested solutions below. There are always multiple methods to solve problems.

1) The pattern in this sequence is add 5 subtract 2. The three numbers that fill in the blanks would be 16, 14, and 19.

Their sum is 16 + 14 + 19 = 49

49

2) It is less expensive to buy the 3 pack so she must purchase as many of those as possible. \$100 ÷ \$13.57 = 7. These cost 13.57 x 7 = 94.99 which leaves 5.01 change and that is not enough for the \$8 shipping cost. So she must go back to the six 3 packs and work from there. \$13.57 x 6 = \$81.42. If she adds the \$8 shipping cost it is \$89.42 and that leaves her \$100 - 89.42 = \$10.58 to spend. She can buy 2 more cans at \$5.19 each or \$10.38 and that will leave her .20¢. She can buy a total of 20 cans.

20

3) The tricky part about this problem is to notice that one cut makes two pieces. So to make 12 logs, it will only take 11 cuts. If it takes 3 minutes to make one cut, then it will take (3 min X 11 cuts) or 33 minutes to cut the 12 logs.

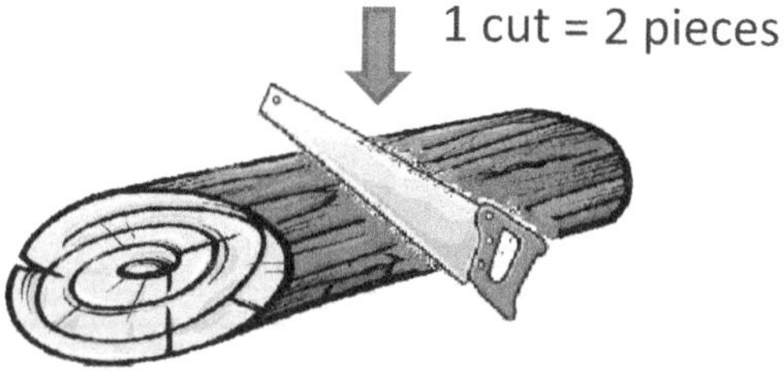

33

PERENNIAL MATH — TEST #2

December 2014 Rookie Level Solutions

4) One way to solve this is to make a chart. Start with half tricycles (10) and half bicycles (11). This makes the total wheels equal to 52. So, we need to decrease the number of tricycles until you reach 48 wheels. There are 15 bicycles.

Starting Point (→ 10)

Tricycles	10	9	8	7	6
Bicycles	11	12	13	14	15
Total Seats	21	21	21	21	21
Wheels	52	51	50	49	48

← Keep Total Seats Constant

15

5) The first step would be to see what shapes are the same on both sides of the scale and eliminate those from the left side. So if we eliminate a triangle, circle, and square from the left, we have 3 stars, 2 triangles, 2 squares, and 1 circle remaining. Since a triangle equals 7 stars, we can count the 2 triangles as 14 stars. A square equals 3 stars, so we can count the 2 squares as 6 stars. A circle equals 2 stars, so we can count those 2 stars. So we would need to add 3 stars + 14 stars + 6 stars + 2 stars = 25 stars

25

6) The Johnson family members each brought a gift for everyone in the family. There are 10 people in the family so each person bought a gift for all the other 9 people. 10 people x 9 gifts each = 90 gifts. 90 gifts x $10 = $900

900

Name________________________________

PERENNIAL MATH

January 2015 Rookie Level

TEST #3

1) Pedro gave Carmen a number. He told her to add 5 to the number and double the sum. Her answer was 40. What number did Pedro give Carmen?

2) Consecutive numbers are counting numbers that follow in order such as 1, 2, 3, and so forth. Find three consecutive numbers that add up to 60. What is the largest of the three numbers?

3) The library charges 15 cents a day fine for the first three days and then 5 cents per day after the third day. Meredith paid a $1 fine. How many days was her book overdue?

4) Mrs. Thompson has 36 students in her class. If there are four girls for every 2 boys, how many girls are there?

5) Find the area of this shape.

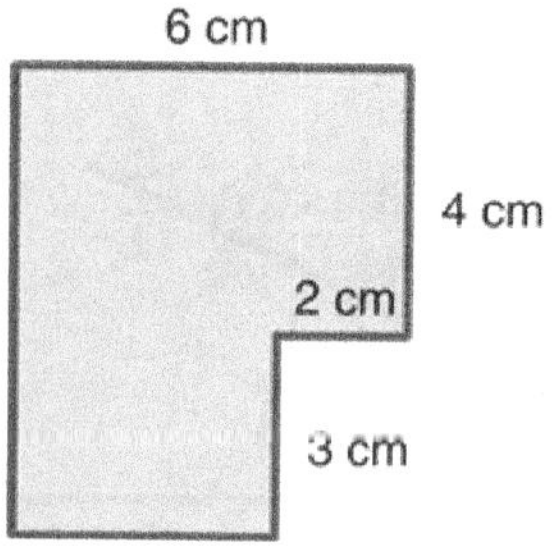

6) Jeremy is between 20 and 60 years old. If you divide his age by 4, there will be no remainder. Next year you will be able to divide his age by 7 without a remainder. How old is Jeremy now?

PERENNIAL MATH

TEST #3

January 2015 Rookie Level Answers

1) Pedro gave Carmen a number. He told her to add 5 to the number and double the sum. Her answer was 40. What number did Pedro give Carmen?	15
2) Consecutive numbers are counting numbers that follow in order such as 1, 2, 3, and so forth. Find three consecutive numbers that add up to 60. What is the largest of the three numbers?	21
3) The library charges 15 cents a day fine for the first three days and then 5 cents per day after the third day. Meredith paid a $1 fine. How many days was her book overdue?	14
4) Mrs. Thompson has 36 students in her class. If there are four girls for every 2 boys, how many girls are there?	24
5) Find the area of this shape. 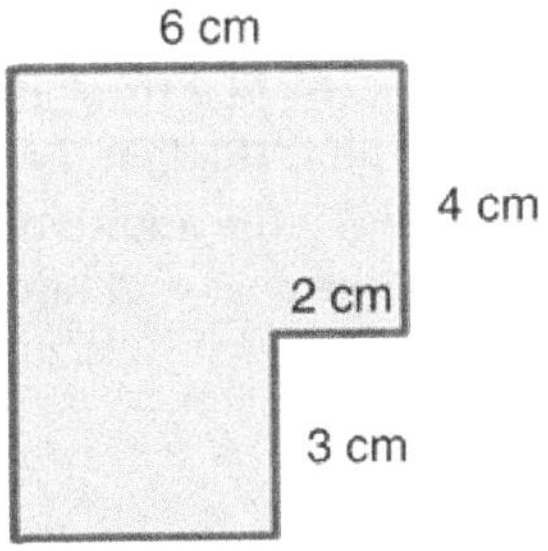	36
6) Jeremy is between 20 and 60 years old. If you divide his age by 4, there will be no remainder. Next year you will be able to divide his age by 7 without a remainder. How old is Jeremy now?	48

PERENNIAL MATH

TEST #3

January 2015 Rookie Level Solutions

We are only providing some suggested solutions below. There are always multiple methods to solve problems.

1) Students can use trial and error to find the answer or work backwards. Start with 40 and divide it in half to get 20. Subtract 5 and you will have 15, the number that Pedro gave to Carmen. You can then work the problem forwards to check your answer.

15

2) Find three consecutive numbers that add up to 60. You can divide 60 by three to get an average (or the middle number). 20 is the average. The three consecutive numbers are 19, 20, and 21. The largest of the three numbers is 21.

21

3) The library charges 15 cents a day fine for the first three days and then 5 cents per day after the third day. Meredith paid a $1 fine. The first three days would be 45 cents leaving an additional 55 cents she paid. If the additional days are 5 cents, then there were (55 ÷ 5) = 11 additional days. Three days + 11 additional days made her book 14 days overdue.

14

PERENNIAL MATH

TEST # 3

January 2015 Rookie Level Solutions

4) Mrs. Thompson has 36 students in her class. There are four girls for every 2 boys. This is a ratio of 4:2 which if you divide the total 36 into 6 equal groups, then 4 groups would be girls and 2 groups would be boys.
36 ÷ 6 groups yields 6 people in each group. So 4 x 6 = 24 girls and 2 x 6 = 12 boys.
Younger students could actually draw this problem out - drawing 4 girls and then 2 boys until they have 36 students.

24

5) Divide the figure into two rectangles and find all missing lengths.

The larger rectangle has an area of:

$4cm \times 7cm = 28cm^2$

The smaller rectangle has an area of:

$4cm \times 2cm = 8cm^2$

If we combine these we will find the total area:

$28cm^2 + 8cm^2 = 36cm^2$

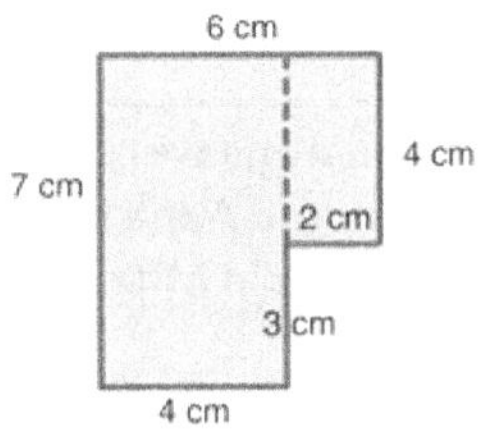

36

6) Since Jeremy is between 20 and 60 years old and there is no remainder if you divide his age by 4, make a chart of possible ages now. Then make another chart for next year's age and it will be a multiple of 7. Jeremy is 48 this year and will be 49 next year.

THIS YEAR	24	28	32	36	40	44	48	52	56
NEXT YEAR	28	35	42	49	56				

48

Name________________________________

PERENNIAL MATH

TEST #4

February 2015 Rookie Level

1) A standard 12 hour clock shows 10 o'clock. What time will it be 100 hours from now?

2) Donny has pennies, nickels, dimes and quarters. How many ways can he combine his coins to make 36 cents?

3) Mr. and Mrs. Perkins have three children. Each of these children are married. Each of those couples have four children. How many total people are in the entire Perkins family?

4) Mitzi, Dale, and Anna went to the jewelry store. Mitzi and Dale had $10 together. Dale and Anna had $8 together. Mitzi and Anna had $14 together. How much did Mitzi have?

5) Using the table below, if the pattern were continued, what number will be in column C in Row 20?

	A	B	C
Row 1	1	2	3
Row 2	6	5	4
Row 3	7	8	9
Row 4	12	11	10

6) There were 72 mailboxes on Bridge Street. One out of every four houses had a yellow mailbox. One out of every six houses had a blue mailbox. How many mailboxes on Bridge Street were not yellow or blue?

PERENNIAL MATH

TEST #4

February 2015 Rookie Level Answers

1) A standard 12 hour clock shows 10 o'clock. What time will it be 100 hours from now?	2
2) Donny has pennies, nickels, dimes and quarters. How many ways can he combine his coins to make 36 cents?	24
3) Mr. and Mrs. Perkins have three children. Each of these children are married. Each of those couples have four children. How many total people are in the entire Perkins family?	20
4) Mitzi, Dale, and Anna went to the jewelry store. Mitzi and Dale had $10 together. Dale and Anna had $8 together. Mitzi and Anna had $14 together. How much did Mitzi have?	8
5) Using the table below, if the pattern were continued, what number will be in column C in Row 20?	58
6) There were 72 mailboxes on Bridge Street. One out of every four houses had a yellow mailbox. One out of every six houses had a blue mailbox. How many mailboxes on Bridge Street were not yellow or blue?	42

Table for question 5:

	A	B	C
Row 1	1	2	3
Row 2	6	5	4
Row 3	7	8	9
Row 4	12	11	10

PERENNIAL MATH

TEST #4

February 2015 Rookie Level Solutions

We are only providing some suggested solutions below. There are always multiple methods to solve problems.

1) If a standard 12 hour clock shows 10 o'clock, it will it be 2 o'clock 100 hours from now. 100 hours divided by 12 is 8 with four hours left over. Four hours past 10 is 2 o'clock.

2

2) Make a chart and notice the patterns that evolve. Start with quarters and work your way down. There are 24 combinations that create 36¢.

Quarters	Dimes	Nickels	Pennies
1	1		1
1		2	1
1		1	6
1			11
	3	1	1
	3		6
	2	3	1
	2	2	6
	2	1	11
	2		16
	1	5	1
	1	4	6
	1	3	11
	1	2	16
	1	1	21
	1		26
		7	1
		6	6
		5	11
		4	16
		3	21
		2	26
		1	31
			36

24

3) Draw the problem out. There are 2 grandparents, 6 parents, and 12 grandchildren. This is 20 people.

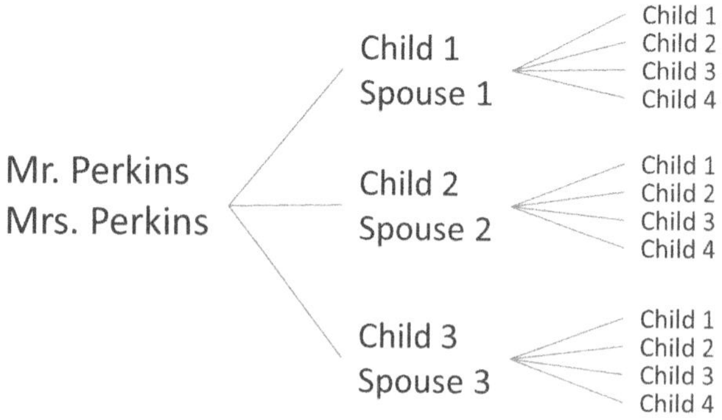

20

PERENNIAL MATH

TEST #4

February 2015 Rookie Level Solutions

4) Mitzi, Dale, and Anna went to the jewelry store. Mitzi and Dale had $10 together. Dale and Anna had $8 together. Mitzi and Anna had $14 together. Students could use trial and error or algebra as shown here.

$$M + D = 10$$
$$D + A = 8 \quad \text{\textit{Add the three equations together}}$$
$$M + A = 14$$
$$2M + 2D + 2A = 32 \quad \text{\textit{Divide in half}}$$
$$M + (D + A) = 16 \quad \text{\textit{Solve for M by substituting known D + A = 8 into equation}}$$
$$M + (8) = 16 \quad \text{\textit{Subtract 8 from both sides}}$$
$$M = 8$$

8

5) The numbers are sequential and just wrap around row to row. Students could draw out the whole chart or they might notice the pattern in just Column C. This pattern is +1 (between Row 1 and 2), then +5 (between Row 2 and 3). Then this pattern is repeated.

The number in Column C in the 20th row would be 58.

	A	B	C
Row 1	1	2	3
Row 2	6	5	4
Row 3	7	8	9
Row 4	12	11	10
Row 5	13	14	15
Row 6	18	17	16
Row 7	19	20	21
Row 8	24	23	22
Row 9	25	26	27
Row 10	30	29	28
Row 11	31	32	33
Row 12	36	35	34
Row 13	37	38	39
Row 14	42	41	40
Row 15	43	44	45
Row 16	48	47	46
Row 17	49	50	51
Row 18	54	53	52
Row 19	55	56	57
Row 20	60	59	58

Column C Pattern
+1
+5

58

6) There were 72 mailboxes on Bridge Street. One out of every four houses had a yellow mailbox. One out of every six houses had a blue mailbox. 72 ÷ 4 = 18 yellow mailboxes. 72 ÷ 6 = 12 blue mailboxes. 72 - (18 + 12) = 42 mailboxes on Bridge Street that were not yellow or blue.

42

Name________________________________

PERENNIAL MATH

TEST #1

November 2015 Rookie Level

1) In a certain code, FIDDLE is written as 825569 and TONE is written as 6479. How is DEFINE written in that code?

2) The Dodgers are one of seven teams in their division. If they play all the other teams 17 games each, how many games do the Dodgers play for the season?

3) A rectangle has a perimeter of 60cm and its length is twice as long as its width. What is the area of this rectangle?

4) Paula took nine pieces of her candy out for herself and then evenly divided the rest among her seven friends. Each friend received thirteen pieces. How many pieces did she Paula start with?

5) How many rectangles of any size are in this shape?

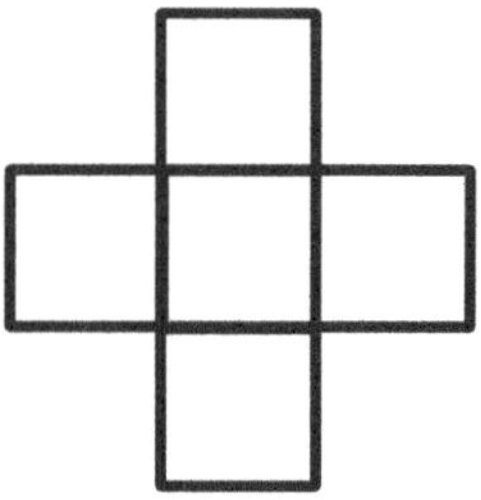

6) Peyton has 20 red balls, and he has 9 fewer yellow balls than red balls. Peyton also has 2 more green balls than red balls. He also has 3 less blue balls than green balls. How many balls does Peyton have in all?

PERENNIAL MATH

TEST # 1

November 2015 Rookie Level Answers

1) In a certain code, FIDDLE is written as 825569 and TONE is written as 6479. How is DEFINE written in that code?	598279
2) The Dodgers are one of seven teams in their division. If they play all the other teams 17 games each, how many games do the Dodgers play for the season?	102
3) A rectangle has a perimeter of 60cm and its length is twice as long as its width. What is the area of this rectangle?	200
4) Paula took nine pieces of her candy out for herself and then evenly divided the rest among her seven friends. Each friend received thirteen pieces. How many pieces did she Paula start with?	100
5) How many rectangles of any size are in this shape? 	11
6) Peyton has 20 red balls, and he has 9 fewer yellow balls than red balls. Peyton also has 2 more green balls than red balls. He also has 3 less blue balls than green balls. How many balls does Peyton have in all?	72

PERENNIAL MATH

TEST # 1

November 2015 Rookie Level Solutions

We are only providing some suggested solutions below . There are always multiple methods to solve problems.

1) In a certain code, FIDDLE is written as 825569 and TONE is written as 6479. How is DEFINE written in that code? Use the numbers and translate DEFINE to 598279.

598279

2) Since the Dodgers are one of seven teams in their division and they play all the other teams 17 games each they would play 6 teams x 17 games = 102 games for the season.

102

3) Since the rectangle has a perimeter of 60cm and its length is twice as long as its width, we can use algebra to represent the width as w and the length is $2w$. Therefore the perimeter can be written as follows,
$2(w + 2w) = 60$. Divide both sides by 2 and $(w + 2w) = 30$. Now you can divide both sides by 3, since $3w = 30$, and determine that $w = 10$. *The width is 10cm and the length is 20cm. The area of the rectangle is*
length x width, $20cm \times 10cm = 200cm^2$

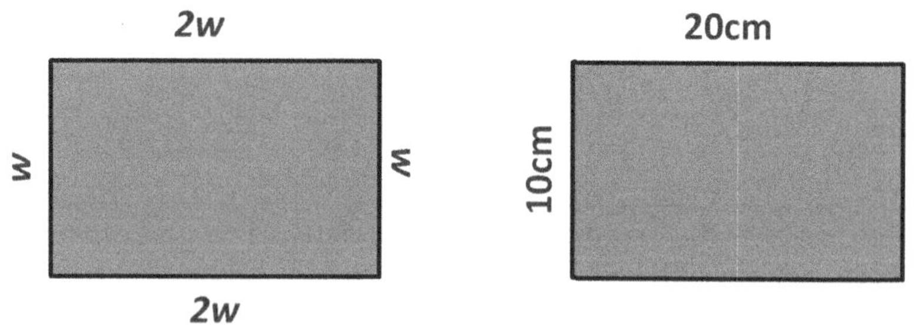

200

PERENNIAL MATH — TEST #1

November 2015 Rookie Level Solutions

4) Paula took nine (9) pieces of her candy out for herself. Then evenly divided the rest among her seven friends who received thirteen pieces each which is 7 x 13 = 91. Add her 9 to her friends 91 and you have:
9 + 91 = 100

100

5) Make a chart to keep up with the type of rectangle and count each type going across first and then down. There are 11 rectangles.

Shapes made with:	Number of that type
1 shape	5
2 shapes	4
3 shapes	2
Total	11

11

6) Make a chart to determine how many of each type of ball and then add them up.

Color	Description	Formula	Total
Red	20 red balls	20	20
Yellow	9 fewer yellow than red balls	20 – 9	11
Green	2 more green than red balls	20 + 2	22
Blue	3 less blue than green balls	22 – 3	19
		TOTAL BALLS	72

72

Name________________________________

PERENNIAL MATH

TEST #2

December 2015 Rookie Level

1) Starting with no money, Haley saved \$49 in March, \$37 in April, and \$18 in May. Then she spent \$15 on music and \$73 on shoes. How much money does Haley have left?	
2) Lori started reading a book that has 257 pages. She read the first 25 pages on Monday night, 30 pages on Tuesday night, and 35 pages on Wednesday night. How many pages does she have left to read?	
3) Barry sold half of his comic books and then bought 8 more. He now has 20. How many did he begin with?	
4) Together Tom and Trevor have 30 cards. Tom gave Trevor 6 of his cards, then they both had the same number of cards. How many cards did Trevor have to start with?	
5) Using all of the numbers 1 to 6 only once, place each number in the circles so that the sum along each side of the triangle equals 12. What is the sum of the three circles on each corner? 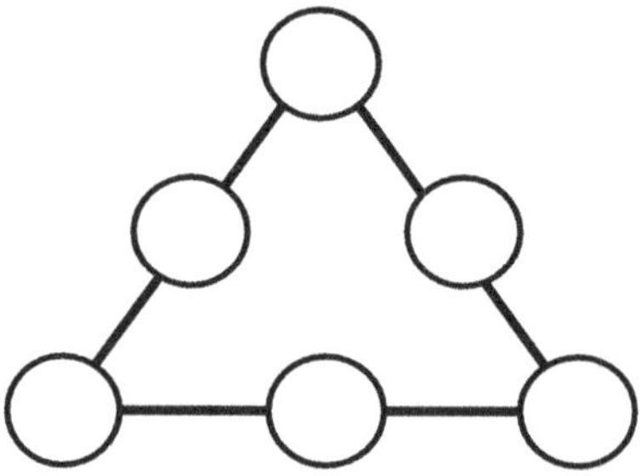	
6) Seven boys formed a chess club and agreed that each boy would play every person only once for a tournament. How many total games were played altogether in the tournament?	

PERENNIAL MATH

TEST #2

December 2015 Rookie Level Answers

1) Starting with no money, Haley saved $49 in March, $37 in April, and $18 in May. Then she spent $15 on music and $73 on shoes. How much money does Haley have left?	16
2) Lori started reading a book that has 257 pages. She read the first 25 pages on Monday night, 30 pages on Tuesday night, and 35 pages on Wednesday night. How many pages does she have left to read?	167
3) Barry sold half of his comic books and then bought 8 more. He now has 20. How many did he begin with?	24
4) Together Tom and Trevor have 30 cards. Tom gave Trevor 6 of his cards, then they both had the same number of cards. How many cards did Trevor have to start with?	9
5) Using all of the numbers 1 to 6 only once, place each number in the circles so that the sum along each side of the triangle equals 12. What is the sum of the three circles on each corner? 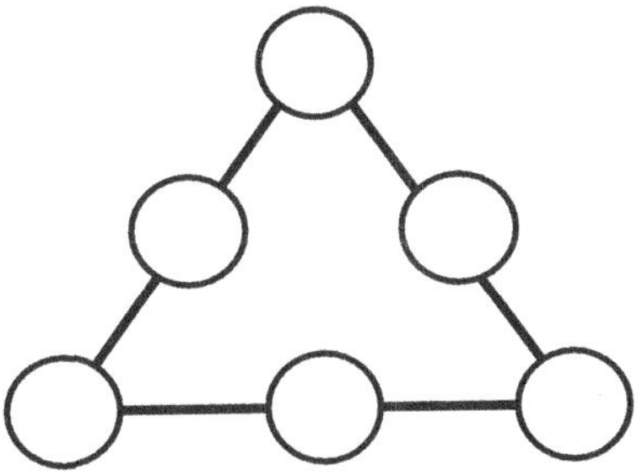	15
6) Seven boys formed a chess club and agreed that each boy would play every person only once for a tournament. How many total games were played altogether in the tournament?	21

PERENNIAL MATH

TEST # 2

December 2015 Rookie Level Solutions

We are only providing some suggested solutions below. There are always multiple methods to solve problems.

1) Haley saved $49 in March, $37 in April, and $18 in May which is $104 total.
Then she spent $15 on music and $73 on shoes which is $88.
Subtract $104 - $88 to find how much money Haley has left.
She has $16

16

2) Lori has already read 25 + 30 + 35 = 90 pages of the 257 page book.
Subtract 257 – 90 = 167 pages left to read.

167

3) Barry sold half of his comic books and then bought 8 more. He now has 20. To find how many comic books he began with work backwards. Twenty minus 8 is 12. Multiply 12 by 2 and he had 24 comic books to start with. Work it forwards to check your work. 24 comic books divided by 2 is 12. Add 8 comic books to the 12 and he has 20. Barry started with 24 comic books.

24

PERENNIAL MATH TEST #2

December 2015 Rookie Level Solutions

4) Together Tom and Trevor have 30 cards. Tom gave Trevor 6 of his cards, then they both had the same number of cards. If the total of the two boys cards is 30 and someone gains six and the other person loses 6 then, half of 30 is 15. So subtract 6 from 15 and add 6 to 15 to get what each boy started with. 15 + 6 = 21, 15 – 6 = 9. Tom started with 21 and Trevor started with 9 cards.
Check this by subtracting Tom's amount 21 - 6 = 15 and adding 6 cards to Trevor's amount 9 + 6 = 15.

9

5) Use logic and look for a pattern. Make a list of the three-number combinations with a sum of 12:

6 + 5 + 1
6 + 4 + 2
5 + 4 + 3

Notice that 6, 5, and 4 are all used twice in the sums.
Arrange them in the corners.
Then the sum of the corners is 6 + 5 + 4 = 15.

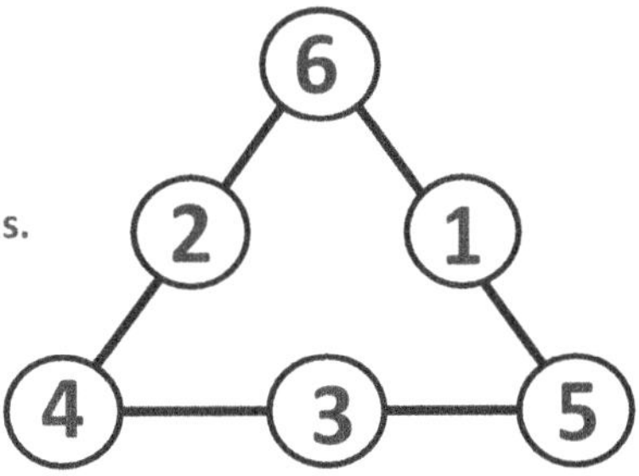

15

6) Seven boys formed a chess club and agreed that each boy would play every person only once for a tournament. Make a chart and list each boy and the match up. You will notice that the pattern is 6 + 5 + 4 + 3 + 2 + 1 = 21. Twenty-one total games were played altogether in the tournament.

Boy 1	Boy 2	Boy 3	Boy 4	Boy 5	Boy 6
Boy 2	Boy 3	Boy 4	Boy 5	Boy 6	Boy 7
Boy 3	Boy 4	Boy 5	Boy 6	Boy 7	1 game
Boy 4	Boy 5	Boy 6	Boy 7	2 games	
Boy 5	Boy 6	Boy 7	3 games		
Boy 6	Boy 7	4 games			
Boy 7	5 games				
6 games					

Boy 2 has already played Boy 1 so don't count him...

21

Name________________________________

PERENNIAL MATH

TEST #3

January 2016 Rookie Level

1) Marcie ran a mile in 10 minutes and 40 seconds. Roslyn ran it in 9 minutes and 51 seconds. How many fewer seconds did Roslyn run the mile in?

2) Mickey wants to ride the swings, the roller coaster, and the spider each twice. The swings costs 9 tickets, the roller coaster costs 8 tickets and the spider costs 7 tickets. Mickey has 5 tickets. How many more tickets should he buy?

3) David is 10 years old. His mother is 4 times his age. After how many years will his mother's age be twice David's age?

4) Bill needs to cut a wooden rod into 5 pieces. It takes 3 minutes to make 1 cut. How many minutes does it take him to cut the rod into 5 pieces if he is cutting at the same rate?

5) Study the first row and find the mathematical pattern. Apply this to the second row to find the missing numbers. What is the sum of the two missing numbers?

5	8	12	15	19
	21	25		32

6) There are 27 students in Mr. Palmer's class. Mr. Palmer decided he would surprise the students with a pizza party! He wanted to make sure each student got at least three slices of pizza. If a medium pizza comes with eight slices, how many medium pizzas did he have to order?

PERENNIAL MATH

TEST # 3

January 2016 Rookie Level Answers

1) Marcie ran a mile in 10 minutes and 40 seconds. Roslyn ran it in 9 minutes and 51 seconds. How many fewer seconds did Roslyn run the mile in?

49

2) Mickey wants to ride the swings, the roller coaster, and the spider each twice. The swings costs 9 tickets, the roller coaster costs 8 tickets and the spider costs 7 tickets. Mickey has 5 tickets. How many more tickets should he buy?

43

3) David is 10 years old. His mother is 4 times his age. After how many years will his mother's age be twice David's age?

20

4) Bill needs to cut a wooden rod into 5 pieces. It takes 3 minutes to make 1 cut. How many minutes does it take him to cut the rod into 5 pieces if he is cutting at the same rate?

12

5) Study the first row and find the mathematical pattern. Apply this to the second row to find the missing numbers. What is the sum of the two missing numbers?

5	8	12	15	19
	21	25		32

46

6) There are 27 students in Mr. Palmer's class. Mr. Palmer decided he would surprise the students with a pizza party! He wanted to make sure each student got at least three slices of pizza. If a medium pizza comes with eight slices, how many medium pizzas did he have to order?

11

PERENNIAL MATH

TEST # 3

January 2016 Rookie Level Solutions

We are only providing some suggested solutions below. There are always multiple methods to solve problems.

1) First convert both runners' times to seconds and then subtract to find the difference. 10 min and 40 sec = 640 secs. 9 min and 51 sec = 591 secs. 640 seconds - 591 seconds = 49 seconds.

49

2)Mickey wants to ride each ride twice. So add up the cost of each ride and multiply it by 2 to find out how many tickets it costs: 9 + 8 + 7 = 24. To ride twice it is 24 tickets x 2 = 48 tickets. He has 5 tickets. So he needs 48 – 5 tickets = 43 tickets.

43

3) David is 10 years old. His mother is 4 times his age. After how many years will his mother's age be twice David's age? David is 10 and his mother is 40. Use this information and make a table to see when his mother will be two times his age. When David is 30 his mother will be 60 which is twice his age. That will be in 20 years.

David	Mother
10	40
11	41
12	42
13	43
14	44
15	45
16	46
17	47
18	48
19	49
20	50
21	51
22	52
23	53
24	54
25	55
26	56
27	57
28	58
29	59
30	60

20

PERENNIAL MATH

TEST #3

January 2016 Rookie Level Solutions

4) Bill needs to cut a wooden rod into 5 pieces. It takes 3 minutes to make 1 cut. How many minutes does it take him to complete the cuts? To cut the rod into 5 pieces-he only has to make 4 cuts. 4 cuts x 3 minutes per cut is 12 minutes.

12

5) Study the first row and find the mathematical pattern. Apply this to the second row to find the missing numbers. What is the sum of the two missing numbers? The pattern is + 3 + 4 + 3 + 4. The missing numbers on the second row are 18 and 28. Their sum is 18 + 28 = 46.

5	8	12	15	19
18	21	25	28	32

46

6) There are 27 students and Mr. Palmer wanted to make sure each student got three slices of pizza. Multiply 27 x 3 to find out how many pieces he will need. He will need 81 pieces. If a medium pizza comes with eight slices, he will divide 81 by 8 to see how many pizzas he needs to order. He will need to order 11 pizzas to make sure everyone gets at least 3 pieces.

11

Name________________________________

PERENNIAL MATH

TEST #4

February 2016 Rookie Level

1) Bob counted 19 bicycles and 27 cars in the garage at the building complex. How many wheels were there in the garage?

2) Callie bought 17 boxes of cat food and 13 boxes of dog food. Each box of cat food contained 7 cans, and each box of dog food contained 3 cans. How many more cans of cat food than dog food did Callie buy?

3) There are 25 students in Mrs. Berry's class. 16 students are football fans, 18 students are baseball fans, 10 students are fans of both sports. How many students are not fans of either sport?

4) Amy opened a bank account in May with a $5 deposit into the account. Amy saved $47 in June, $19 in July, and $38 in August. Then Amy spent $23 on school supplies and $54 on new clothes. Her grandmother sent her $25 in the mail. How much money does Amy have now?

5) This 3 x 3 grid is made up of a border of eight squares surrounding one square. A 3 x 4 grid would be made up of a border of ten squares surrounding two squares. How many squares would be in the border of a 3 x 20 grid?

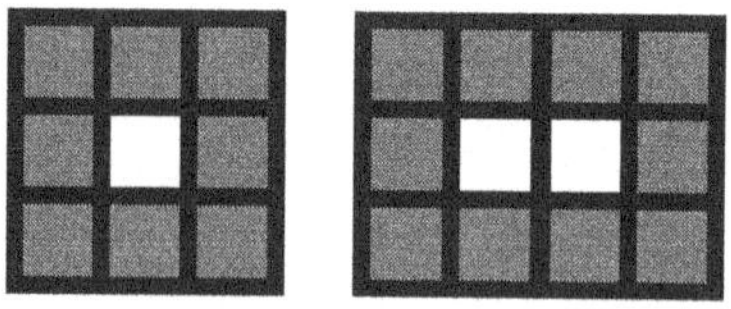

6) What is the sum of the digits of all the even numbers from 0 to 30?

PERENNIAL MATH TEST #4

February 2016 Rookie Level Answers

1) Bob counted 19 bicycles and 27 cars in the garage at the building complex. How many wheels were there in the garage?	146
2) Callie bought 17 boxes of cat food and 13 boxes of dog food. Each box of cat food contained 7 cans, and each box of dog food contained 3 cans. How many more cans of cat food than dog food did Callie buy?	80
3) There are 25 students in Mrs. Berry's class. 16 students are football fans, 18 students are baseball fans, 10 students are fans of both sports. How many students are not fans of either sport?	1
4) Amy opened a bank account in May with a $5 deposit into the account. Amy saved $47 in June, $19 in July, and $38 in August. Then Amy spent $23 on school supplies and $54 on new clothes. Her grandmother sent her $25 in the mail. How much money does Amy have now?	57
5) This 3 x 3 grid is made up of a border of eight squares surrounding one square. A 3 x 4 grid would be made up of a border of ten squares surrounding two squares. How many squares would be in the border of a 3 x 20 grid? 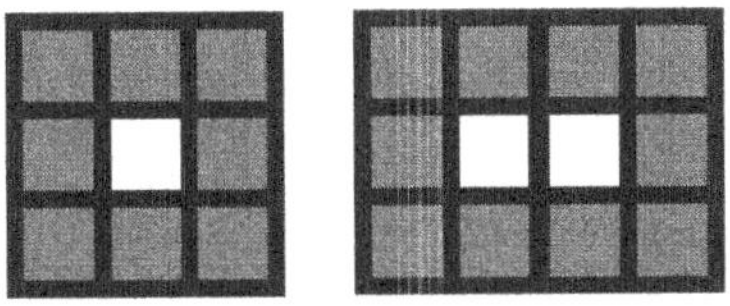	42
6) What is the sum of the digits of all the even numbers from 0 to 30?	78

PERENNIAL MATH

TEST #4

February 2016 Rookie Level Solutions

We are only providing some suggested solutions below. There are always multiple methods to solve problems.

1) Bob counted 19 bicycles and 27 cars in the garage at the building complex.
(19 x 2 = 38 wheels) plus (27 x 4 = 108 wheels). 38 wheels + 108 wheels = 146 wheels.

146

2) Callie bought 17 boxes of cat food and 13 boxes of dog food. Each box of cat food contained 7 cans, and each box of dog food contained 3 cans. To find out how many more cans of cat food Callie bought, do the calculations to determine how many cans of each and subtract.
17 x 7 = 119 cans of cat food. 13 x 3 = 39 cans of dog food. 119 – 39 = 80

80

3) To find our how many students are not fans of either sport use a Venn diagram. Divide the regions and solve for the region outside by subtracting all the inside regions from the total amount of students. Football only (6) + Both (10) + Baseball only (8) = 24 students. 25 total students – 24 combined sports fans = 1 student who is not a football or baseball fan.

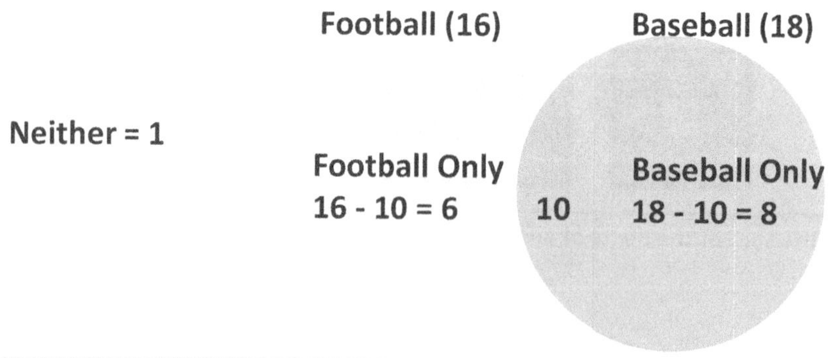

1

PERENNIAL MATH

TEST #4

February 2016 Rookie Level Solutions

4) Amy's starting balance of $5 + ($47 + $19 + $38) = $109. She spent ($23 + $54) =$77. Subtract what she spent from her bank account. ($109 - $77) =$32. Finally add the $25 her grandmother sent her. ($32 + $25) = $57 Amy now has $57.

57

5) This 3 x 3 grid is made up of a border of eight squares surrounding one square. A 3 x 4 grid would be made up of a border of ten squares surrounding two squares. To find out how many squares would be in the border of a 3 x 20 grid you can draw out the figure and count up the squares or you can look for a pattern. The second number, which we will name n + 2, where n stands for what number or place the drawing is in the sequence. For example 3 x <u>3</u> is the <u>first</u> item so 3 x ((1) +2) or 3 x <u>4</u> is the <u>second</u> item or 3 x ((2) + 2). So 3 x 20 would be the 18th item in the sequence. To find the answer the equation is 3 x (n + 2) – n. So you could deduce that 3 x 20 – 18 = 42 squares in the border of the 3 x 20 grid.

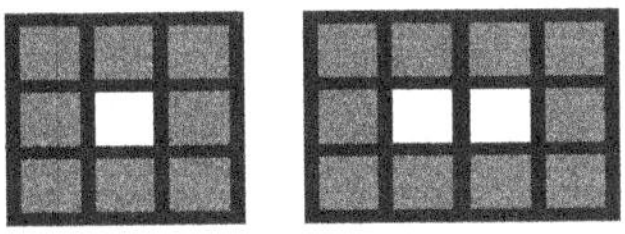

42

6) To find the sum of the digits of all the even numbers from 0 to 30, list the numbers out and then add up each digit. REMEMER you are adding up digits, so 12 is equal to 1 + 2.

2 + 4 + 6 + 8 + 1 + 0 + 1 + 2 + 1 + 4 + 1 + 6 + 1 + 8 + 2 + 0 + 2 + 2 + 2 + 4 + 2 + 6 + 2 + 8 + 3 + 0 = 78.

78

Name________________________________

PERENNIAL MATH

TEST #1

November 2016 Rookie Level

1) There are 17 boys and 8 girls in Mrs. Stuart's class. Yesterday 5 boys and 3 girls were absent. So there were ______more boys than girls in the class yesterday.	
2) Gigi is four years younger than her brother Barry. Barry is 26 years younger than his mother. Barry's father is 39, two years older than Barry's mother. How old is Gigi?	
3) A farmer has some chickens and rabbits. There are sixteen heads and thirty-eight feet in the pen. How many chickens does he have?	
4) In this subtraction problem, the letter A stands for the same digit everywhere it appears. What digit does A stand for? $$\begin{array}{r} 432 \\ -\ \underline{AA} \\ 3AA \end{array}$$	
5) There are 120 pieces of candy in a bowl. Mark takes a third of them. Then, Carol takes a fourth of what remains. Finally, Chuck takes a fifth of what remains after Mark and Carol took their shares. After Chuck leaves, Leslie eats the rest of the candy in the bowl. How many pieces of candy did Leslie eat?	
6) Room A and Room B have 58 students in total. 4 students are transferred from Room A to Room B. Now the two rooms have the same number of students. How many students were in Room A before the transfer?	

PERENNIAL MATH

TEST # 1

November 2016 Rookie Level Answers

1) There are 17 boys and 8 girls in Mrs. Stuart's class. Yesterday 5 boys and 3 girls were absent. So there were ______more boys than girls in the class yesterday.	7
2) Gigi is four years younger than her brother Barry. Barry is 26 years younger than his mother. Barry's father is 39, two years older than Barry's mother. How old is Gigi?	7
3) A farmer has some chickens and rabbits. There are sixteen heads and thirty-eight feet in the pen. How many chickens does he have?	13
4) In this subtraction problem, the letter A stands for the same digit everywhere it appears. What digit does A stand for? $\begin{array}{r} 432 \\ -\ \ AA \\ \hline 3AA \end{array}$	6
5) There are 120 pieces of candy in a bowl. Mark takes a third of them. Then, Carol takes a fourth of what remains. Finally, Chuck takes a fifth of what remains after Mark and Carol took their shares. After Chuck leaves, Leslie eats the rest of the candy in the bowl. How many pieces of candy did Leslie eat?	48
6) Room A and Room B have 58 students in total. 4 students are transferred from Room A to Room B. Now the two rooms have the same number of students. How many students were in Room A before the transfer?	33

PERENNIAL MATH — TEST # 1

November 2016 Rookie Level Solutions

1) There are 17 boys and 8 girls in Mrs. Stuart's class. Yesterday 5 boys and 3 girls were absent. So to find out how many more boys than girls were in the class yesterday subtract the absent boys and girls from the original amount of each and then subtract boys minus girls.

17	8
-5	-3
12 boys	5 girls

12 boys – 5 girls = 7 more boys than girls

7

2) Work backwards. Barry's father is 39, two years older than Barry's mother . So, she has to be 37. Barry is 26 years younger than his mother. So Barry is 37- 26 = 11. Gigi is four years younger than her brother Barry. So Gigi is 11-4 = 7 years old.

7

3) Make a table to compare the total feet. You will notice that if you start with half chickens and half rabbits that you need to decrease the number of rabbits. Also, when you add a chicken and take away a rabbit, the number of feet decreases by 2. There are 13 chickens in the pen.

Chickens	8	9	10	11	12	13
Rabbits	8	7	6	5	4	3
Chicken Feet	16	18	20	22	24	26
Rabbit Feet	32	28	24	20	16	12
Total Heads	16	16	16	16	16	16
Total Feet	48	46	44	42	40	38

Decreases by 2 when a rabbit is added

13

PERENNIAL MATH

TEST # 1

November 2016 Rookie Level Solutions

4) You may use trial and error to solve the problem or you can use logic to decide what A can be. If you notice that 2-A = A in the units column, then the only possibilities would be 1 or 6. However, A cannot be 1 because 3-A = A in the tens column and 3-1 does not equal 1. Therefore, A must equal 6.

$$\begin{array}{r} 432 \\ -\ \underline{AA} \\ 3AA \end{array} \qquad \begin{array}{r} 432 \\ -\ \underline{66} \\ 366 \end{array}$$

6

5) Make a table to keep up with each process:

	Starting Amount	Fraction	Subtract from starting amount	Ending Amount
Mark takes 1/3	120	120 ÷ 3 = 40	120 - 40 = 80	80
Carol takes 1/4	80	80 ÷ 4 = 20	80 - 20 = 60	60
Chuck takes 1/5	60	60 ÷ 5 = 12	60 - 12 = 48	48
Leslie eats remainder	48		48 - 48 = 0	

Leslie eats the remaining 48 pieces.

48

6) Room A and Room B have 58 students in total. 4 students are transferred from Room A to Room B to make the two rooms have the same number of students. So if the total is 58, then both rooms would have to end up with 29 students each. So the Room B had 29 – 4 = 25 and Room A had 29 + 4 = 33 to begin with. To check this 25 + 33 = 58.

33

Name________________________________

PERENNIAL MATH

December 2016 Rookie Level

TEST #2

1) If you add two numbers the sum is 20. If you subtract the smaller number from the larger number, the answer is 2. What is the product of the numbers?	
2) A painter had to paint the house number on houses from 100 to 199. How many ones did he have to paint?	
3) Rex spent 95 cents. He bought a ruler for 49 cents, an eraser for 19 cents and three pencils. How many cents did he pay for each pencil?	
4) How many rectangles of any size are in the picture below? 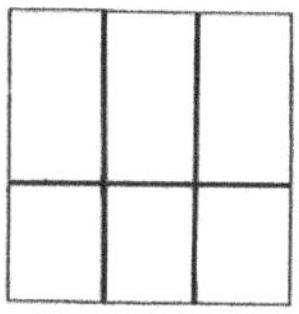	
5) Marlene has 5 baskets that hold 21 balls total. No basket has the same amount of balls in it and only one basket can have an odd number of balls. What is the largest number of balls that she can put in one of the baskets?	
6) A soccer team went to a tournament. The coach brought 48 packs of crackers which allowed each player to get 3 packs of crackers. If there were 4 more boys than girls on the team, how many girls were on the team?	

PERENNIAL MATH

TEST #2

December 2016 Rookie Level Answers

1) If you add two numbers the sum is 20. If you subtract the smaller number from the larger number, the answer is 2. What is the product of the numbers?	99
2) A painter had to paint the house number on houses from 100 to 199. How many ones did he have to paint?	120
3) Rex spent 95 cents. He bought a ruler for 49 cents, an eraser for 19 cents and three pencils. How many cents did he pay for each pencil?	9
4) How many rectangles of any size are in the picture below? 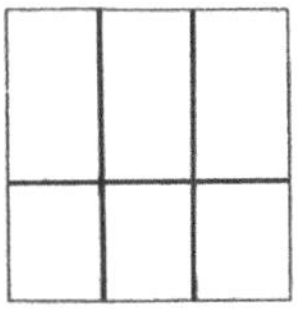	18
5) Marlene has 5 baskets that hold 21 balls total. No basket has the same amount of balls in it and only one basket can have an odd number of balls. What is the largest number of balls that she can put in one of the baskets?	8 or 14
6) A soccer team went to a tournament. The coach brought 48 packs of crackers which allowed each player to get 3 packs of crackers. If there were 4 more boys than girls on the team, how many girls were on the team?	6

PERENNIAL MATH

TEST # 2

December 2016 Rookie Level Solutions

1) If you add two numbers the sum is 20. If you subtract the smaller number from the larger number, the answer is 2. What is the product of the numbers?
First solve for numbers that add up to 20 but have a difference of 2. The two numbers are 11 and 9. Now solve for their product.
11 x 9 = 99

99

2) A painter had to paint the house number on houses from 100 to 199. To solve for how many ones did he have to paint first add up the number of ones that would appear in the hundreds digit, then the tens digit, and then the ones digit.

	Numbers	How many times 1 appears
Hundreds digit	100 -199	100
Tens digit	110, 111, 112, 113, 114, 115, 116, 117, 118, 119	10
Ones digit	101,111, 121, 131, 141, 151, 161, 171, 181, 191	10
Total		120

120

3) Rex spent 95 cents. He bought a ruler for 49 cents, an eraser for 19 cents and three pencils. To solve for how much he paid for each pencil subtract his ruler and eraser purchases and then divide the remainder by three.

$95 - (49 + 19) = 95 - 68 = 27$

$27 \div 3 = 9.$

He paid 9 cents for each pencil.

9

PERENNIAL MATH TEST #2

December 2016 Rookie Level Solutions

4) Make a chart to count the different types of rectangles in the picture: There are 18 total rectangles in this picture. (Don't forget to count down and across)

How many rectangles make the shape (type)	How many rectangles of each type
1	6
2	7
3	2
4	2
5	0
6	1
Total	18

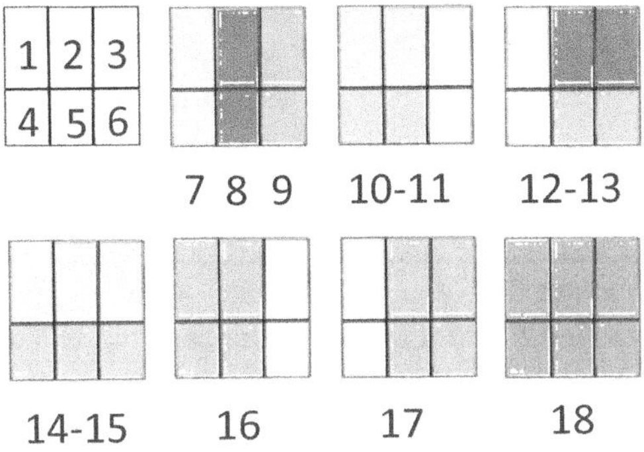

18

5) Marlene has 5 baskets that hold 21 balls total. No basket has the same amount of balls in it and only one basket has an odd number of balls. To find the largest number of balls that she can put in one of the baskets just put the lowest number of balls in the first 4 baskets and make 1 be the number in the odd basket.

1 + 2 + 4 + 6 + ?

The largest basket holds 21 - 13 = 8 balls.
The question is ambiguous meaning that it doesn't specify that a basket cannot have zero balls, so the answer could be 14. (14, 4, 2, 1, 0).

8 or 14

6) The coach brought 48 packs of crackers which allowed each player to get 3 packs of crackers. Since there were 4 more boys than girls on the team, to solve for how many girls were on the team, first find the number of team members there were by diving the crackers by 3 packs per player. 48 ÷ 3 = 16. Since there were 4 more boys than girls, there would be 10 boys and 6 girls.

6

Name________________________________

PERENNIAL MATH

TEST #3

January 2017 Rookie Level

1) What is the sum of the first 10 odd numbers greater than 25?

2) Tommy is 13 years younger than his brother. If the sum of their ages is 27, how old is his brother?

3) Take a look at this sequence 100, 98, 94, 88, 80...
If 100 is the first number in the sequence, what will the 10th number be?

4) Use the information from the scale to solve for the values of each item on the scale. How much would three stars weigh?

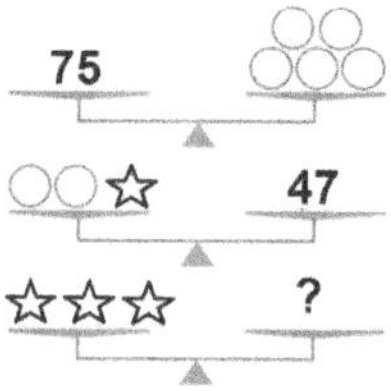

5) The area of a rectangle is 100 in^2. One side is 4 inches. What is the perimeter of the rectangle?

6) In the video game club each member played each of the other members only one time. Over 25 games were played. What is the least possible number of members in the club?

PERENNIAL MATH

TEST # 3

January 2017 Rookie Level Answers

1) What is the sum of the first 10 odd numbers greater than 25?	360
2) Tommy is 13 years younger than his brother. If the sum of their ages is 27, how old is his brother?	20
3) Take a look at this sequence 100, 98, 94, 88, 80... If 100 is the first number in the sequence, what will the 10th number be?	10
4) Use the information from each scale to solve for the value that belongs in the last scale. How much do three stars weigh? 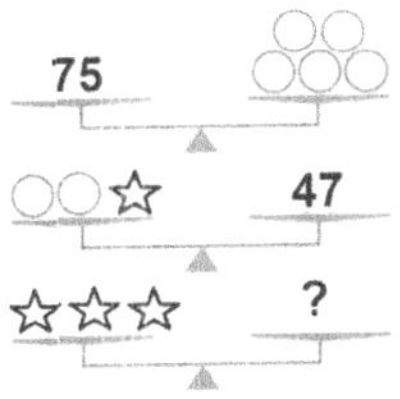	51
5) The area of a rectangle is 100 in^2. One side is 4 inches. What is the perimeter of the rectangle?	58
6) In the video game club each member played each of the other members only one time. Over 25 games were played. What is the least possible number of members in the club?	8

PERENNIAL MATH

TEST #3

January 2017 Rookie Level Solutions

1) List the 10 numbers and use smart addition to group the digits in the units (ones) column:

The units column adds up to 50. The tens column add up to 310. 50 + 310 = 360

27
29
31
33
35
37
39
41
43
45

360

2) Tommy is 13 years younger than his brother. Since the sum of their ages is 27, to find out how old his brother is you can use trial and error or make a table to compare their ages. His brother is 20.

Tommy	Brother (Tommy + 13)	Sum
1	14	15
2	15	17
3	16	19
4	17	21
5	18	23
6	19	25
7	20	27

20

3) Look for a pattern in the sequence. You will notice that the numbers decrease by 2, 4, 6, 8. So each number is decreasing by 2 more than the previous decrease. Make a chart. You will see that the 10th number in the sequence is 10.

Position in Sequence	Number	Subtraction
1	100	
2	98	-2
3	94	-4
4	88	-6
5	80	-8
6	70	-10
7	58	-12
8	44	-14
9	28	-16
10	10	-18

10

PERENNIAL MATH

TEST # 3

January 2017 Rookie Level Solutions

4) Using the information from each scale you can solve for the value that belongs in the last scale. Five circles weigh 75 so one circle equals 75/5 = 15. Using this new information then 2 circles plus 1 star weighs 47. So 2 Circles weigh 30 and the star would weigh 17. Now you can solve for how much the three stars weigh. 3 x 17 = 51

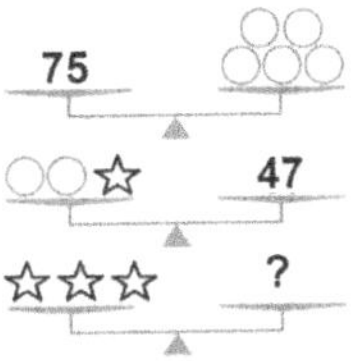

51

5) The area of a rectangle is 100 in². One side is 4 inches. To find the perimeter of the rectangle we must solve for the dimensions of the rectangle first. Since one side equals 4 inches, the other side must equal 25 inches because area is equal to length times width. Now we can use these dimensions to solve for the perimeter. 2 (25) + 2 (4) = 58 inches.

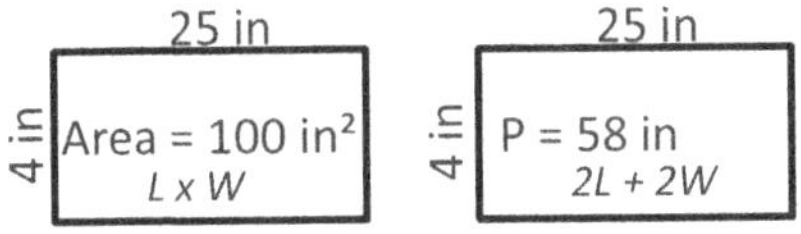

58

6) If you started with 2 players there would be 1 game played. With three players there would be 2 more games played for a total of 3 games. With four players there would be three more games played. See the pattern in the table. There has to be at least 8 players because 7 players would only equal 21 games played.

2 Players	Add Player 3	Add Player 4	Add Player 5	Add Player 6	Add Player 7	Add Player 8
Player 1 v Player 2	Player 3 v Player 1	Player 4 v Player 1	Player 5 v Player 1	Player 6 v Player 1	Player 7 v Player 1	Player 8 v Player 1
1 game	Player 3 v Player 2	Player 4 v Player 2	Player 5 v Player 2	Player 6 v Player 2	Player 7 v Player 2	Player 8 v Player 2
	2 games	Player 4 v Player 3	Player 5 v Player 3	Player 6 v Player 3	Player 7 v Player 3	Player 8 v Player 3
		3 games	Player 5 v Player 4	Player 6 v Player 4	Player 7 v Player 4	Player 8 v Player 4
			4 games	Player 6 v Player 5	Player 7 v Player 5	Player 8 v Player 5
				5 games	Player 7 v Player 6	Player 8 v Player 6
					6 games	Player 8 v Player 7
						7 games

<u>8 Players</u>

1 + 2 + 3 + 4 + 5 + 6 + 7 + 8 = 28 games played

8

Name_______________________________

PERENNIAL MATH

TEST #4

February 2017 Rookie Level

1) Fran made a mistake on a math problem. She added 9 to her answer instead of multiplying by 9. Fran's answer was 17. The correct answer to her problem is _____.

2) Lance took a math test with 40 problems. He earned 5 points for each problem he got right, no points if he left it blank and lost 3 points for each wrong answer. He had 19 wrong answers and left 6 questions blank. What was his score on the test?

3) How many hours are in 2 weeks, 2 and a half days?

4) How many rectangles are in the diagram below?

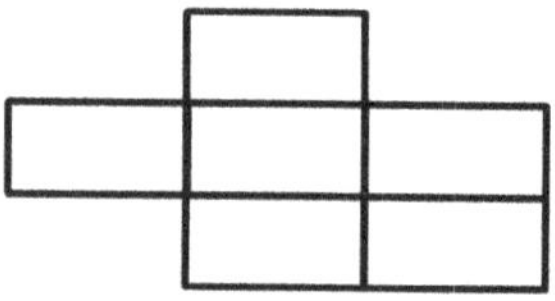

5) John wants to donate videos from his two book shelves. The first book shelf has 4 shelves with 26 videos on each shelf. The second book shelf has 4 shelves with 24 videos on each shelf. Every video is the same size so John packs the videos evenly into eight boxes. How many videos will there be in each box?

6) What is the ones digit of $19 \times 19 \times 19 \times 19$?

PERENNIAL MATH TEST #4

February 2017 Rookie Level Answers

1) Fran made a mistake on a math problem. She added 9 to her answer instead of multiplying by 9. Fran's answer was 17. The correct answer to her problem is _____.	72
2) Lance took a math test with 40 problems. He earned 5 points for each problem he got right, no points if he left it blank and lost 3 points for each wrong answer. He had 19 wrong answers and left 6 questions blank. What was his score on the test?	18
3) How many hours are in 2 weeks, 2 and a half days?	396
4) How many rectangles are in the diagram below? 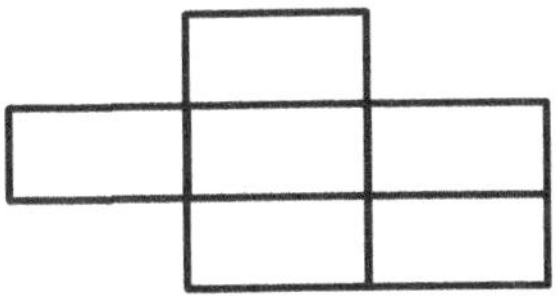	15
5) John wants to donate videos from his two book shelves. The first book shelf has 4 shelves with 26 videos on each shelf. The second book shelf has 4 shelves with 24 videos on each shelf. Every video is the same size so John packs the videos evenly into eight boxes. How many videos will there be in each box?	25
6) What is the ones digit of 19 x 19 x 19 x 19?	1

PERENNIAL MATH

TEST #4

February 2017 Rookie Level Solutions

1) Fran added 9 to her answer instead of multiplying by 9. Her answer was 17. So work backwards to find her number: 17 – 9 = 8. So the correct answer to her problem is 8 x 9 = 72

72

2) Lance took a math test with 40 problems. He earned 5 points for each problem he got right, no points if he left it blank and lost 3 points for each wrong answer. He had 19 wrong answers and left 6 questions blank.

Subtract to find out how many problems he got right. 40 – (19 + 6) = 15. Now plug in the totals for each type of answer:

Correct answers: 15 x 5 = 75
Wrong answers: 19 x 3 = 57
75 – 57 = 18

18

3) To find out how many hours are in 2 weeks, 2 and a half days first multiply to find out how many hours are in 16 days which is 2 weeks and 2 days. 16 days times 24 hours per day = 384. Now add the 12 hours that are a half day. 384 + 12 = 396 hours.

396

PERENNIAL MATH

TEST #4

February 2017 Rookie Level Solutions

4) To find out how many rectangles are in the diagram devise a table and count the types of rectangles going across and down.

How many rectangles make the shape (type)	How many rectangles of each type
1	6
2	6
3	2
4	1
Total	15

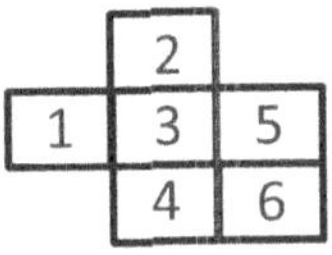

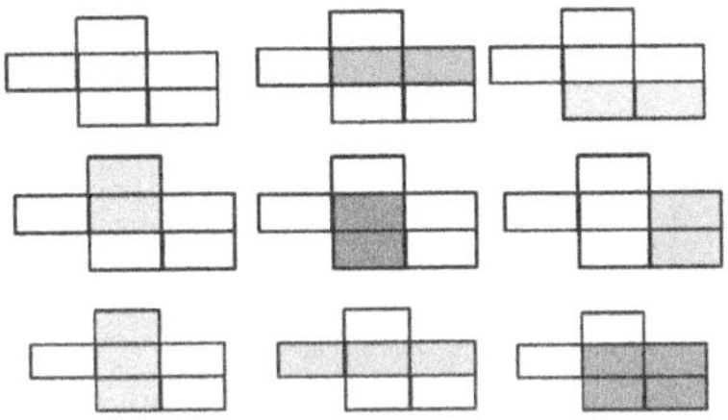

15

5) The first book shelf has 4 shelves with 26 videos on each shelf or 4 x 26 = 104 videos. The second book shelf has 4 shelves with 24 videos on each shelf or 4 x 24 = 96 videos. He has a total of 104 + 96 = 200 videos. He packs them equally into 8 boxes so 200 ÷ 8 = 25. There are 25 videos in each box.

25

6) To find the ones digit of (19 x 19 x 19 x 19) you don't have to do the entire problem. However if you do, the product is 130,321. But you can simply multiply 9 x 9 x 9 x 9 and get the ones digit. Or group (9x9) x (9 x 9) which is 81 x 81. There you will see that the ones digit is 1.
Note: A Multiplication pattern arrives with the ones digit when multiplying nines: 1, 9,1,9,1,9...

1

Made in United States
North Haven, CT
04 November 2026

21763948R00029